NAOMI'S ROAD

JOY KOGAWA
NAOMI'S ROAD

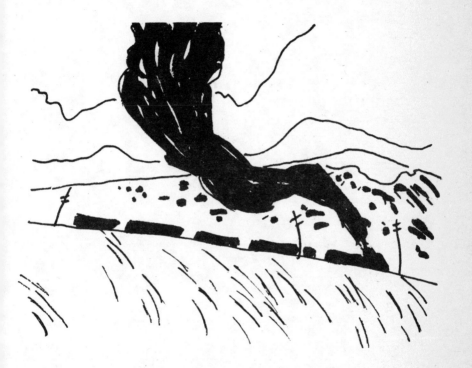

Drawings by MATT GOULD

Toronto
OXFORD UNIVERSITY PRESS

Oxford University Press, 70 Wynford Drive, Don Mills, Ontario M3C 1J9

Toronto   Oxford   New York
Delhi   Bombay   Calcutta   Madras   Karachi   Kuala Lumpur
Singapore   Hong Kong   Tokyo   Nairobi   Dar es Salaam
Cape Town   Melbourne   Auckland   Madrid

and associated companies in
Berlin   Ibadan

This book is for my new sister
Michiko Asami

Canadian Cataloguing in Publication Data

Kogawa, Joy.
  Naomi's road

An adaptation for children of the author's novel Obasan.
ISBN 0-19-540547-1

1. Japanese Canadians — Evacuation and relocation,
1942–1945 — Juvenile fiction.*   I. Gould, Matt.
II. Title.

PS8521.044N36 1986   jC813'.54   C86-093856-5
PZ7.K63Na 1986

A letter from the author

Dear Children,

O Canada! What a vast, beautiful country. Here there are people from all around the world. And along with the Native Peoples, we are all Canadians together.

This little story is told by a Canadian child called Naomi Nakane. She has black hair and lovely Japanese eyes and a face like a valentine. Naomi's story happened in the days before you were born, in the 1940s. In her childhood there was a war going on. Canada and Japan were enemies. How sad that was. Suddenly she had to be ashamed to be Japanese. She did not learn to read or write Japanese and she tried to forget how to speak Japanese. She never used chopsticks with strangers.

It is hard to understand, but Japanese Canadians were treated as enemies at home, even though we were good Canadians. Not one Japanese Canadian was ever found to be a traitor to our country. Yet our cameras and cars, radios and fishing boats were taken away. After that our homes and businesses and farms were also taken and we were sent to live in camps in the mountains. Fathers and older brothers and uncles were made to work building roads in the Rocky Mountains. If you ever drive through these beautiful mountains, you may ride over some roads made by Japanese Canadians.

Naomi's road is a different kind of road. It is the path of her life. If you walk with her a while, you will find the name of a very important road.

1

"Bong, bong, bong . . ." Six, seven, eight. The grandfather clock in the hall says it's almost time to go to bed. Night's shadowy light is soft and quiet.

We're all here in the music room, Daddy and Mama and me and my big eight-year-old brother, Stephen. He's sitting on the piano stool beside Mama, his heels bumping against the piano stool's skinny legs. Mama's long skirt touches the floor. Sometimes her buckled shoes tap against the metal bird claws at the bottom of the piano stool legs. Each claw clutches a green glass ball. Behind Stephen, Daddy is bobbing up and down in time to the music.

"And one, and two, up'n one, and two . . ."

Daddy's long fingers hold a pencil and he makes squiggly marks on Stephen's music books.

I'm wearing my nemaki—a nightie—and sitting in the high-backed wicker chair by the windows. Mama gave me a tea biscuit after my bath tonight. The goldfish in the bowl by the window ledge look like they want a piece.

Mama waves to me to sit beside her. "Come," she

says, but I'm going to stay with the goldfish. We listen as Mama sings one of my kindergarten songs. Her voice is yasashi—soft and tender. Mama is always yasashi.

> "How did you, Miss Daffodilly,
> Get your pretty dress?
> Is it made of gold and sunshine?
> Yes, child, yes."

The wicker chair is big enough for me to curl up in. I make myself round as a peach while Mama sings. Daddy comes and nibbles my biscuit. He pretends he's a monkey at the zoo. He makes such silly faces as he nibbles.

"Sleepy?" Mama asks me when she finishes singing. "Have you had the biscuit?"

I nod my head. I'm not really sleepy, but I want to go to bed to hear the story.

"Goodnight," we say to Daddy and Stephen and the goldfish and off we go.

First we walk through the dark living room where Stephen and I sometimes play with his red-coated lead soldiers. Then into the dining room and across to my bedroom with its long, white lacy curtains.

Past the curtains are the branches of the peach tree right outside my window. Once when my window was open a robin came and stood on the ledge and didn't fly away. It looked right at me. Above my bed I have a picture of a little green bird in a green tree. The sky is green too.

Mama pulls back the blue patchwork quilt and I climb into bed. I can smell the powder that she pats on her face and neck with her powder puff.

"What story shall I tell tonight?" she asks.

"Momotaro. Tell me Momotaro," I say to Mama almost every night. I love the story of Momotaro. It's my favorite ever since I was a baby.

Sometimes Mama lies down on the bed beside me and her sweet perfume smell is close against my face.

"Once upon a time, a long time ago," she begins, "there was an old old man and an old old woman who lived in a little cottage in a forest." They're very very dear but so lonely, Mama says, because they have no children. Little children are more special and dear than anything you can think of. One day, when the old woman is washing clothes, she finds a huge golden peach rumbling and tumbling down the stream.

"Ah, and what do you suppose is in the peach?" Mama asks.

When I was little I used to clap my hands and hide my face in the pillow every time Mama asked this.

"Momotaro!" I'd cry out. "Momotaro!"

Even now though I'm bigger, I feel happy just thinking of the beautiful baby peach boy hidden in the middle of a giant peach.

4

I hug Mama and she sings the peach boy song. "Momotaro, Momotaro, Momotaro-san."

I'd love to find a peach baby. I'd love to have a dear old grandfather and a dear old grandmother. Most of all though, I want to be a child forever and forever.

But children grow up.

"That's the way it goes," Stephen says.

2

The small green peaches on the peach tree outside my window are the size of Stephen's marbles at first. Daddy says we'll eat them when they're as big as my bouncing ball.

One afternoon, when some of the peaches are almost ready to pick, I'm playing with my dolls by the window. I have a teddy and a mouse, a nurse doll and my favorite Japanese baby doll. The doll has tiny red lips with two little teeth. Her hair is just like mine—short with straight bangs. I have a blue and white tea set with little spoons as thin as toothpicks and a tin stove with a tin kettle full of water. The mouse is singing its mouse song with me when Ralph, the big boy who lives down the alley, comes to play. He's looking for Stephen.

''What're you doing?'' he asks. He sits down on the floor and picks up my stove. ''Want your kettle to boil?'' Ralph takes a box of matches from his pocket.

I've never made a fire before. Ralph takes out a match and shows me how to swish it against the side of the box. The fire is so sudden and so hot that I drop the match. Then—surprise! A swift curl of fire runs along the edge of the lace curtain. The fire leaps and flies upwards. It's fast as a bird flying out of a bush.

"Oh oh!" Ralph shouts. He sounds afraid. "Look what you did, Naomi!" He runs out of the room. "You'll get a spanking."

I stand for a moment watching the curtain burning. My mother and father never spank me or Stephen. The fire catches the other curtain and the flames rush to the window top.

Mama is downstairs. She'll know what to do. I run down and call her.

"Mama!"

She comes quickly. Bits of burning curtains are dropping through the air. She doesn't shout like Ralph did but runs back and forth with water from the kitchen. And there, on the floor, the dots of fire turn into soggy puddles.

When it's over, we sit on my bed and look at the round black spot on the ceiling.

"What happened?" Mama asks quietly.

The burnt match on the floor is shrivelled and black. It crumbles as I pick it up.

"Dangerous, isn't it?" Mama says softly. Her voice is yasashi. I'm glad she's not afraid like Ralph was. Because Mama isn't afraid, I'm not afraid either.

"A match is safe if you know how to blow it out,"

Mama says. Everything's safe where Mama is. I sit close to her, the safest place in the world.

But one day, Mama's packing to go away. My great-grandmother in Japan is sick.

"Come," Mama says when she sees me standing in the doorway watching.

"Can I go with you?" I ask as I help her pack.

Mama takes her crystal necklace off and puts it around my neck. "Another time," she says, straightening the necklace.

"Will you be back for my birthday?" I ask. I'm going to be six soon.

Mama nods. "Obasan will take care of you. And you'll take care of Obasan, won't you?" she says. "Obasan will make lots of onigiri." Obasan is my aunt. When she visits she brings sticky onigiri rice balls with the salty red plum in the middle.

The day Mama goes away, it's bright and sunshiny. We're down by the sea where the big boats are—Daddy and Stephen and Uncle and Obasan and me. Streamers and streamers and streamers are everywhere. The pink and yellow and blue and green ribbons of paper twirl and sway through the noisy air. It's like a giant maypole dance. Once, when I was very little, I ran into a maypole dance to find Mama but I got caught in the streamers. Today Mama is nowhere. Daddy lifts me in his arms and points to the boat. "See?" he asks. "Can you see?"

Stephen is running around picking up some unused rolls of streamers. He stuffs them into Daddy's pockets.

I can't see Mama. All the streamers are in the way. All the people are in the way.

"Mama," I call out. I want to shout louder but shouting isn't polite.

In all the noise I can't hear her answering me. Instead, the boat blasts its giant whistle and suddenly all the people's hands are as windy and wild as my peach tree branches in a storm.

"Goodbye, Mama," Stephen shouts. "Goodbye!"

I wish I could hide inside Mama's coat, like I did when I was little. I look down at the white woolly flowers she sewed on the bottom of my blue wool dress. One of them is coming undone. I pull the flower off for Mama to fix. She even put some perfume on the flowers so they would smell real.

When we get home, I put the woolly flower under my pillow so I can still smell Mama when I go to bed. Stephen comes to say goodnight.

"Here," he says and gives me three of his streamer rolls.

Daddy says Obasan is going to sleep on a cot in my room until Mama comes home.

In the morning, I'm surprised to see Obasan's long black braid hanging down her back. I've only seen her hair in a bun at the back of her head. Stephen and I watch as she puts hairpins in her bun. Hairpins, hairpins, she has a hundred hairpins.

"What has long legs and crooked thighs, no head and no eyes?" Stephen asks. It's a hairpin riddle.

But Obasan doesn't understand riddles. She doesn't understand English very well. She smiles at Stephen anyway.

Obasan is soft and gentle like Mama. Her velvet dressing gown and her quilt are soft too. Sometimes I curl up in her cot because her quilt is so fluffy.

3

I don't understand why Mama is taking so long to come home.

"She'll come back when she can, Little One," Daddy says.

Stephen is getting grumpier and grumpier. "But when is that, Daddy?" Stephen asks.

"I don't know exactly," Daddy says.

"You don't know. I don't know. We don't know. They don't know." Stephen says. "Nobody knows anything." Stephen stomps into the music room and plays loud angry music on the piano.

The questions make everyone unhappy. After a while I stop asking.

At night, I sleep with my Japanese doll and whisper secrets to her. I wish and wish my doll could talk. One of my picture books that Mama used to read is about a doll that came to life.

"How can I make you talk?" I ask my doll. But she never says anything.

I pretend that she has a teeny tiny voice that only I can hear. When I hold her mouth up to Daddy's ear, he smiles. But he says he can't hear her.

"What's she saying, Button Face?" Daddy asks.

"She's asking you when Mama's coming home."

"Oh," Daddy says. Then he stares and stares at the sky. "She can't come home till the war is over," he says quietly.

"What's war?" I ask.

Daddy tells me that war is a terrible terrible thing. It is the worst and saddest thing in the world. People get hurt and learn to be afraid. It's like the time the burning match made the fire in my room. War is more dangerous even than that.

One afternoon Stephen comes running home from his music lessons. His glasses and his violin are broken. There are black wet crying marks on his face. Obasan is hanging clothes on the line when she sees him.

"What happened?" she asks softly. Her voice is yasashi, like Mama's.

Stephen doesn't answer. Obasan takes his broken violin and they go into the kitchen. She wipes his face. Stephen is cross because I'm watching. So I take my doll and go downstairs. We hide under Daddy's cot in his study. My doll cries because Stephen was crying.

"There's a war. That's why Stephen got hurt," I tell the doll.

My doll jumps up and down angrily on the floor. "War is stupid!" she shouts. Now my doll can talk. But she can't say words very well. "Toop it! Toop it!

13

Toop it!" she says. The doll is so angry she breaks her hand. Then she cries and cries because she is hurt and wants Mama to come home and fix her.

After awhile the doll lies down on her back and stares up at the bottom of the bed. All over the mattress, there are white cotton tufts like bunny tails.

"I wish Mama was here," the doll says. "Then we would be safe as bunnies."

"Go to sleep, Dolly," I tell her. "I'll take care of you."

Later at night, she wakes me up because she's frightened. I'm frightened too. The night light in the hall is out. The street lights are out. Such darkness! This is what Stephen calls a "black-out." The whole city is hiding. If an enemy in an airplane sees us he might drop a bomb. The bomb would make a huge fire and burn the house. Then what would we do?

I'm so afraid I can hardly move. I want to find Daddy. He's playing the piano softly in the music room. I feel my way in the darkness till I find him.

"Couldn't you sleep, my Little One?" he asks.

I climb into his lap and hug him tight.

He sings the daffodilly song. And then he sings "The mountain and the squirrel had a quarrel," a funny song Stephen learned at school. Daddy is trying to make me laugh. But I don't feel like laughing now when the whole world is dark.

4

"You're a big boy, Stephen," Daddy says one day. "Take good care of your sister and practise your piano."

We're playing with Stephen's cars and trucks and soldiers on the deep blue Indian rug in the living room. The lead soldiers are marching on the zigzag border. We pretend it's a winding road. Soldiers aren't as much fun as dolls. All they can do is fight and fall down dead.

Daddy kneels beside us. He has to go away, he says. And Uncle has already gone. "Be good, children. Listen to Obasan," he says. "Say your prayers every day."

That's the last thing Daddy tells us before he goes away. And then one day soon, Obasan says we're going away too.

"Where are we going?" Stephen asks.

"On a holiday," she says. "Imagine! Mountains! And a train!"

"Will we be gone for a long time?"

"Perhaps," Obasan says.

"We're going on a holiday! We're going on a holiday!" I tell my doll.

"Oh," says the doll and does a tap dance on the floor.

But on the day we leave, my doll feels afraid. She doesn't like the noise of all the people at the train station. The small children are holding their mothers' hands and legs and skirts. Some of them are holding their dolls like I am. Many of the children look scared. They don't answer me when I say hello to them.

None of the white children from our street are here. None of the white children from Stephen's school are here either.

The train is full of strangers. It smells of oil and soot and orange peels. If you stand up you almost fall over, as if you were on a rocking chair. The soot on the window ledge jiggles and jumps like little black flies.

A few seats in front, there's a tiny red-faced baby. The baby's eyes are closed and the mouth is squinched shut small as a button. If I lean out farther, I can see the tiny pink fist. It's squashed like a marshmallow against the baby's cheek.

"Go and see the new baby," Obasan says. She gives me an orange to take. But I feel shy. There are too many strangers. And Mama and Daddy aren't here.

Obasan finds a towel and some apples and oranges and takes them to the baby's mother. She bumps from side to side as she goes. The baby's mother bows

deeply. She almost folds herself in half over the baby.

Close to our seats there's a humpbacked little old woman. Her back is round as a church bell. She bounces off the seat as Obasan comes back. The old woman is so short. When she stands she's shorter than when she was sitting. She holds the train seat and bends forward.

"Something for the baby," she says. She begins to take off her white flannel underskirt.

"Ah, ah, Grandmother," Obasan says gently.

"It is clean," the old woman says. "Last night it was washed."

Obasan holds her in the rock-rock of the train. They sway together back and forth. The old woman is careful not to let the underskirt touch the floor.

"For a diaper," the old woman says. She folds the underskirt into a neat square. Her fingers are stiff and curled. They look like the driftwood you find on the beach.

Obasan bows and takes the present. She puts it on the young mother's lap. Their heads bob like birds as they talk. I hold my doll up so she can watch them.

Outside the train window, the trees are zipping past. What would it be like, I wonder, to be lost in the woods. I would have to walk and walk and walk. But I might never find the way home. At night there'd be wolves and bears. Maybe there would even be snakes.

What if I climbed up a tree and couldn't get down again?

I stop looking at the trees and bring out my toys to play with. I have a red, white, and blue ball, and a Mickey Mouse that can walk by itself down a slope. I ask Stephen if he wants to play, but he stares out the window. He's doing piano exercises with his fingers on his knees.

After a while my doll feels sleepy. I put her to bed on a blanket on Obasan's lap. There's enough room for my head beside the doll. Obasan's face is quiet and calm. Her hand taps my back as I fall asleep.

"Nen, nen," Obasan sings. It's a lullaby Mama used to sing to me when I was a baby. "Nen, nen."

5

The train coughs and shudders as it stops. What place is this? There are mountains and mountains everywhere. People are crowding out of the train.

I hold onto Obasan's skirt. If I let go, I'll be lost. She's carrying two suitcases. Such jumping and bumping and boxes and bags. It's even noisier than the train station in Vancouver. But there are no streetcars or street lights here.

"Look, it's Slocan," Stephen says. He points to a sign on the train station. Behind the station, the tops of the mountains are white and purple. "Bears are up there," Stephen says, "and eagles and lynxes."

I'm busy following Obasan. She's found a man with a wheelbarrow and she's looking for some boxes in a huge pile of luggage. A boy in a gray suit has a kitten. The kitten seems to be mewing but there's so much noise I can't hear it.

And then we're walking along a road following a minister and the man with the wheelbarrow. I met the minister before in Vancouver. He has a round face and round glasses. He used to visit us in our house.

After we've walked awhile, we come to a rickety wooden bridge. Below us the water jumps and skips along, bouncing over the rocks. On the bank of the stream, a crow is hop-hopping on skinny legs as stiff as lead soldiers' legs. I want my doll to talk to the crow.

"Where's my doll?" I ask, calling ahead to Obasan.

Obasan stops and turns around. She looks at the luggage on the wheelbarrow.

Her voice sounds worried. "Ara!" she says.

She lifts pieces of luggage one by one from the wheelbarrow. My doll is lost!

The man pushing the wheelbarrow comes and squats down in front of me. "We'll find the doll," he says. He slaps his knees as if he's already found her. His round face is full of crinkly laugh lines.

"Did we leave her on the train?" I ask.

"We'll find her," the man says.

I want to cry because my doll is lost. But Obasan says not to worry. She takes my hand.

Stephen is ahead of us. The woods are so thick I can't see him.

"Wait! Stephen!" I call, running to catch up.

He's staring and staring. "See that?" he says, pointing. I can't see anything except trees and trees and more trees. Beneath us, the ground is soft and bumpy. There are fat pine cones and little acorn hats.

Ferns are spread open like green fans.

"Can't you see it?" Stephen says. He's impatient with me.

We walk a few more steps and there, hidden in the woods, is a small gray hut.

"That's where we're going to live," Stephen says.

It looks like a giant toadstool. It's surrounded by tall weeds. Is it a real house, I wonder.

Stephen clumps up the porch steps and pushes open the front door. It scrapes along the floor. How gray everything is. There's a dead dried bumble-bee on the window sill. Dusty newspapers cover the walls instead of wallpaper. Everything looks gray. I've never seen such a dusty little house. Maybe it's the home of the three bears. But there's no porridge waiting in a great big bowl or a middle-sized bowl or a wee little bowl. There's only the dead bee by the gray window, and the weeds outside that look as if they want to come in.

6

Every morning I wake up in a narrow bunk bed near the stove. I wish and wish we could go home. I don't want to be in this house of the bears with newspaper walls. I want to be with Mama and Daddy and my doll in our real house. I want to be in my own room where the picture bird sings above my bed. And the real bird sings in the peach tree outside my bedroom window. But no matter how hard I wish, we don't go home.

The house is so crowded we can barely move around. In one small room there are two beds. One is for Obasan. The other is for a long-faced woman called Nomura-obasan. She's not well, Obasan says, and we must take care of her.

Daddy's sick too, Stephen says. His letters are from a hospital somewhere in the woods.

"When is he coming here?" I ask Obasan one night. We're sitting at the table after supper. The coal-oil lamp is on. "When will he get better?"

Nobody answers me. Nobody knows.

Stephen is practising his pieces on a folding cardboard piano Daddy made. "The world is beautiful as long as there is music," Daddy wrote. "Keep the world beautiful, Stephen. If you listen hard you can hear all the notes."

Sometimes Stephen and I pretend we're at home again in our music room and the cardboard piano is real. We play guessing games and I have to guess which songs he's playing. Even if I'm older now, I like singing the kindergarten songs the best.

Obasan is washing the supper dishes. She fills the basin from the water bucket by the stove. "Plip" says the dipper and "szt szt" goes the water as it spills on the hot stove. The box beside the stove is full of logs and kindling wood. Obasan and Stephen chop the logs outside on a stump.

Behind the house there's a path that goes up the mountain. If we climbed all the way we'd reach the sky. On our way up Stephen and I find tart red strawberries the size of shirt buttons. And there are gooseberries, shiny and round as marbles. We find floppy dark mushrooms too, growing on dead trees. Obasan will know if they're safe to eat. In early spring curly fiddleheads poke out of the ground. They look like green question marks. We fill our jam pails and bring them all home to Obasan.

From a high rocky ledge past a waterfall, we can see the world. Far below is the silvery river. And further away, rows and rows of little houses are tiny as toy blocks. Pencil-thin lines of smoke curl out of chimneys. Hundreds and hundreds of boys and girls like Stephen and me live in the toy block houses. Two families share each house and each family has one room. If you wanted to walk around you'd have to be as small as a doll.

In the spring and summer we all play outside. But then winter comes.

One cold day Stephen and I are playing outside. The minister and another man are carrying a cot through the fluffy falling snow.

"For Uncle," the minister says when Stephen points to the cot.

"What?" Stephen interrupts excitedly. "Is Uncle coming here?"

When we get home, Obasan nods solemnly. "Yes, Uncle is coming tonight."

"Really?" I ask. "Is Daddy coming too? Can we go home?"

Nomura-obasan shakes her head sadly. "Not yet," she says.

"Come," Obasan says brightly. She wipes her hands on her apron. "There's so much to do. Just think! Uncle is on his way."

We're like elves hopping about all afternoon. Obasan cooks the dried mushrooms and fiddleheads. I make paper decorations and paper baskets for jelly beans.

Even Nomura-obasan tries to help, but her hands are too shaky.

As we work, the snow keeps falling. The fence post looks like it's wearing a tall hat. Stephen puts his hand on the window to melt the frost so he can see. But after a while it gets dark.

At last we hear a stomp stomp outside. Stephen throws the door open and in comes Uncle in a whoosh of snow.

"Uncle!" Stephen cries.

Uncle puts down his wooden box and sack and shakes the snow off his coat. His arms are wide as Papa Bear's. "Hello hello hello," he says as he lifts Stephen up.

Obasan takes off her apron. She folds her hands in front of her. "Welcome home," she says. "You are just in time."

Uncle looks at all the food and the decorations on the table. "Ah," he says, "it must be Christmas."

"You have come such a long way," Nomura-obasan says. She is sitting up in bed and bows forward. Uncle bows as well and they both say, "It is such a long time."

Then he squats in front of me and scratches his head.

"And this big girl. Who can she be?" he asks. He's joking, of course, but I wonder if I've changed. He still looks the same.

He turns to his sack and takes out two wooden flutes. With a whoop, Stephen leaps to Uncle. And then Stephen's fingers are dancing lightly over the smooth wood. At once the room fills with a bright dancing sound. Uncle slaps his knees as Stephen hops around and round the wooden box chairs. Stephen is like a rooster, crowing with his head up high. He plays and plays.

"Oh there will be dancing," Nomura-obasan says, clapping her hands.

"You're just like your father," Uncle says, patting Stephen on the back. "Music all the time."

7

With Uncle here now, the little house changes. Shelves and benches and wallpaper appear.

One calm summer day, Obasan and Uncle are tending the garden. Stephen is drumming on a tub drum and I'm at the lake with Kenji, a boy my age who lives in one of the small block houses. His wobbly glasses bounce up and down his nose when he jumps.

He's paddling around on a log raft. I'd like to go with him but Uncle says I mustn't go on rafts. The water is cool and tickly against my toes.

I'm wearing my green and white bathing suit. Close by is our sand-castle village. We make sand houses with grass-top trees and white-pebble sidewalks. There are twig chimneys, twig bridges, twig people, and one fat twig dog with three legs.

Kenji and I are playing when Rough Lock Bill comes to watch us. Rough Lock Bill is a tall skinny man with hair like seaweed. He sits on the sand and talks to us as we play. I can see his big toe sticking out of his sock.

"What's your name?" he asks me.

"Her name's Naomi," Kenji replies.

"Can't talk?" he asks me again. He hands me a stick. "Here," he says. "Can you print? Print your name."

I brush the wet sand off my hands and take the stick. NAOMI, I print in large letters.

"Aha," Rough Lock Bill says. "Naomi."

He picks up one of the stick people. "Is this you?" he asks me.

I shake my head. After a while he picks up all our stick people and tells us stories. When he gets tired he goes back to his house beside the beach. We can see him rocking in his chair.

Kenji gets tired of our sand village. "Let's go in the water," he says. He jumps up and runs to his raft. The raft wobbles as he gets on.

"Come on, Naomi," he calls. A long pole is in his hand. He pushes the raft closer to the shore. "Climb on," he says. "I won't go far."

The water makes my toes curl up tight. I've never gone into deep water. But I kneel on the wobbly raft.

"Okay," Kenji says. He leans on the pole. The raft scoots out over the water. I lie down on the wet logs. My toes dangle over the edge. There are small gray fish swimming beside us. After a while the water feels colder. When I look up, I can't see the sand village at all.

"It's too far here, Kenji," I say. I feel afraid now. "Let's go back."

"Just one more shove," Kenji says. He leans back, lifting one leg and the pole high in the air. Down goes the pole straight into the water. Then—splash! Kenji tumbles sideways and falls overboard. Whoosh! The cold water sprays over my back, and the raft bounces on the waves. I feel like a cat on a see-saw. There's nowhere to hold. I'm on my hands and knees trying to balance.

Kenji is paddling around and shaking the water out of his hair. He has his glasses in his teeth, like a dog. I can see the pole behind us further out in the lake. Kenji's trying to say something to me through his teeth as he swims to shore. But I can't hear him.

When he gets to the beach he stands up and waves. "Jump!" he shouts.

"I can't," I shout back. "I can't swim."

He holds his hands up to the sides of his head. He can't see me without his glasses. Then slowly, he steps backwards till he's out of the water.

I know he can't help me. He doesn't know how to help. His hands are stiff beside him. I know he's afraid. He turns and runs down the beach. He won't come back.

The raft drifts in the still lake. Last month a boy drowned in this lake. I must decide quickly what to do. If I wait, the raft will go further and further and I'll be lost.

I feel sick and afraid. Then I jump.

Down I go. Down, down, down. Water is in my ears. Water is in my nose. Water is in my head. Water is everywhere. I can't breathe. I can't see. I can't tell where the air is or where the shore is. I splash and gasp and swallow air and water. Again and again I'm whirled around in the choking dizziness. I try to cry out but the sound I make is like an animal growling.

After a horrible time something is pulling me along through the water. I feel as limp as laundry on a line. Suddenly my ears clear. There's a whack whack on my back.

"Okay, okay, I gotcha," a man's voice says.

Between gasps, I'm breathing. I'm breathing and I know I'm safe. Water gushes from my nose and mouth.

Rough Lock Bill places me on my stomach on the sand. He turns my head sideways on his red and blue plaid shirt. I can feel the sand against my cheek. When I move my knees up, more water comes out of my stomach and throat and nose. I wipe my nose on the sleeve and close my eyes.

"That's it," Rough Lock says. He peers at me.

More than anything in the world, I wish Mama was here.

After a while, Rough Lock Bill carries me piggyback to our hut. He tells Stephen what happened and Stephen tells Obasan and Uncle.

"Thank you. Thank you," Uncle says to Rough Lock Bill. "Thank you for saving Naomi."

8

When I'm well again, Obasan tells me I must learn to swim.

"Water is dangerous," she says, "if you cannot swim."

We're in the women's bathhouse in Slocan with mothers and girls and little children. Uncle and Stephen are on the other side of the wall in the men's bathhouse.

Our bathhouse has a long bench by the door where we put our clothes. I'm squatting on the wood-slatted floor, scrubbing Obasan's back. Nomura-obasan is here today. She's so well now that she can come to the bath. She's going to move back to her daughter's family.

"When you learn to walk, you can climb mountains," Obasan says. "And when you learn to swim, you can ride on rafts. Life is for learning many things."

I dip the basin into the big square bath full of hot water. So many girls and women are soaking and chatting in the steamy room. When I rinse all the soap off Obasan, we climb back into the hot water. I feel so drowsy I almost fall asleep. Obasan tells me that there are many things to learn. I must learn to read and print and write. I must learn to add and subtract. I must learn about animals and insects and people and trees. School is a place to learn all these things.

A year has passed since we came to Slocan, and a school has been built. It's half a mile away, where everyone lives in the rows and rows of huts.

The path to school is through the forest. Every school day, Stephen and I carry our lunches and schoolbags. For lunch I have a boiled egg and dandelion greens and Obasan's onigiri rice balls with a salty red plum in the middle. Stephen takes sandwiches and an apple or an orange.

On the way home from school, Stephen and I walk by a big white house. There's a swing in the back yard. A pretty girl about my age lives there. She has light golden hair like Goldilocks. Sometimes, before we reach her yard, we can see her swinging on her swing. Higher and higher she goes, her toes pointing up to the sky.

One day we stand at the fence and watch her.

"Boy," Stephen says. "I bet she'll go right around."

Toys are all over her backyard, like a toy store. There's a doll carriage and a doll house and a doll's tea set on a doll's table. And there are two real live white bunnies hopping in a pen. I wish I could hold the bunnies. They look as fluffy and soft as cotton wool.

The golden-haired girl sees us standing at the fence.

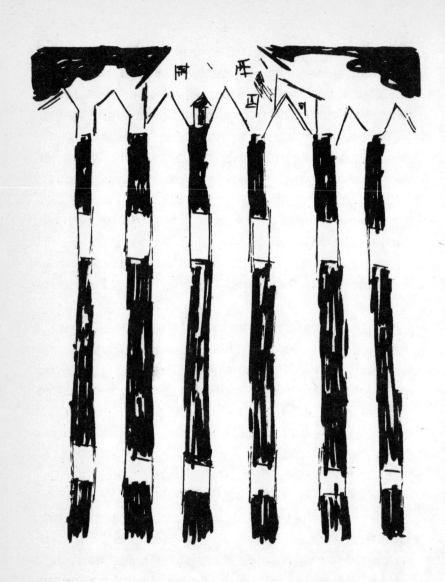

She scrapes her feet on the ground to stop her swinging. Then she jumps off.

"What are you staring at?" she asks. She sounds angry. I want to run away into the trees. She makes a face at us and stomps her feet.

"Go away," she shouts.

What a mean girl. "Come on," I say to Stephen. I start to walk down the path.

But Stephen is angry. He whacks at the grass with his lunch pail.

"Go 'way," the girl shouts again.

"Why should I?" Stephen says. "This is a free country."

"It's not your country," the girl says.

"It is so," Stephen says.

"It is not," she shouts.

A red and white checked curtain in the window behind her moves. There's a woman who also has golden hair, looking out at us. She raps on the window with her knuckles. The girl looks back. The woman is shaking her head.

"I can't play with you," the girl says in a sing-song voice. She points her chin to the sky and turns her head.

I run through the trees, taking a short cut away from the path. The thick pine-needle floor crackles as I go. I can hear Stephen behind me hitting the trees with his lunch pail.

9

I don't like the horrid girl. I don't like walking by her house. I don't like school either. In the morning I don't like having my hair combed. My hair is getting long and Obasan braids it in pigtails. I don't like my pigtails.

But I like climbing the mountains. I like playing with my Mickey Mouse who can walk by itself down a slope. I like reading my grade-three reader and Stephen's grade-five reader. And I like reading the comics in the newspaper.

There are some funny roly-poly comic-strip boys called the Katzenjammer Kids. They play tricks on a mean little rich boy called Rollo. And there's a fuzzy-haired girl with empty-circle eyes called Little Orphan Annie. She is always saved from danger by her Daddy Warbucks. Sometimes I lie in my bunk bed at night pretending I'm Little Orphan Annie being rescued by my Daddy.

Stephen likes to read the comics too. But he also reads the harder parts of the newspaper. He says he has to know what's happening in the war. Uncle and

Stephen talk about the war together while they chop wood.

One day Stephen comes running home with a red, white, and blue Union Jack. It's the same as the flag high up on a pole at school. He holds it high in the air and the flag flaps behind him.

"Where did you get that?" I ask.

"I won it," Stephen says. "I traded it for all my marbles." Back and forth he waves the flag. Then he nails it to a long pole and plants it in a hole at the top of Uncle's rock garden. The flag hangs quietly and peacefully high up in the air.

When Stephen jumps back down again, he stands at attention facing the flag. Then he salutes it.

"We have to sing 'God save the King,'" Stephen says. He makes a trumpet out of his hands. After that we sing "Land of Hope and Glory" and "O Canada." When we are singing "Hearts of Oak," I see the horrid girl walking up the path.

We stop. She stops too. She's staring at us and staring at the flag.

"That's not your flag," she says.

"It is too," Stephen says.

"You stole it," she shouts. "Give it to me."

"It's mine," Stephen shouts back.

"You're going to lose the war," she says in her sing-song voice.

"We will NOT!" Stephen yells so loud I cover my ears and run into the house.

From inside the house Obasan and I listen to Stephen pounding on the tub drum. Obasan's eyes are shut.

After a while Stephen comes in and climbs into his bunk. He lies down and takes the flute from under his pillow. All the songs he can remember, he plays and plays and plays. Even when it's time to sleep he keeps playing. Uncle joins in with the tappity-tappity sounds of spoons on his knees.

"Good music," Uncle says to Stephen.

"Good drumming, Uncle," Stephen replies.

When Nomura-obasan was with us, she used to say, "Music will heal us all." Obasan says it now, quietly, with her eyes closed. Obasan is still praying.

10

Three days pass. It's around noon. I'm playing at the side of the house making a little pond with rocks and flowers and a bowl of water. Obasan is washing clothes in the back with a washboard and tub. All my clothes from Vancouver are too small now and Obasan has added hems and sleeves to make them bigger.

I'm putting a buttercup in the bowl for a pretend lily pad when I hear someone saying "Hi."

The horrid girl and her mother are in the middle of the road. They are both shielding their eyes from the sun as they look up. The Union Jack flaps coolly in the mountain air.

"Hello," the mother says to me. "I can see the flag from the window."

I stand up. I feel shy and want to go to Obasan.

"I'm Mitzi," the girl says. "What's your name?"

"Naomi."

"Hello, Naomi," the mother says. She's smiling.

"Hello."

Mitzi comes up to the fence and leans on it. "Can you come and play at my house?" she asks.

I can hardly believe what she's saying. Will she let me pet her bunnies?

"Ask your mother," Mitzi's mother says.

She doesn't know Obasan is my aunt.

When I ask Obasan, she wipes her hands on her apron and nods. She goes into the house and brings out a bag of cookies.

"For Mitzi," Obasan tells me.

"Oh, thank you," Mitzi's mother says to Obasan when I give them to Mitzi. "Say 'thank you,' Mitzi."

"Can I eat one?" Mitzi asks.

Obasan smiles and Mitzi's mother smiles. It seems to me that the trees and the birds and the sun and the flag and all the creatures in the whole world are smiling right now.

Mitzi skips down the path munching the cookie.

"Come on," she calls to me. I feel too shy to skip but I walk quickly to keep up.

When we come to her yard, Mitzi breaks one cookie into little pieces. She puts them in a little doll's dish on her doll's table.

All afternoon we play together. I cuddle her bunnies. At first they make little jerky jumping movements with their back feet. But afterwards they get used to me. They eat sticks of carrots and pieces of lettuce. Their wriggly noses sniff and sniff. One is called Patsy and the other is called Gruff. Mitzi tells me that when they have babies I can have one. I want to jump up and run home and tell Stephen.

Almost every day after this, Mitzi and I make up games and concerts. We make a playhouse out of blankets in the trees. We make mud pies and pine-needle tea and have tea parties with the dolls. One time when we're playing house, she wears my best bead necklace. She likes it so much, I let her keep it.

Mitzi has three favorite dolls. One has eyes that close with a "click" sound when you lay her on her back. She's a fancy doll in a lacy dress. She has white socks and white shoes and tiny white shoelaces. When you spank her or put her on her stomach, she makes a crying noise.

The second doll is a Raggedy Ann with long pigtails like mine. She was a Christmas present when Mitzi was four. She loves her Raggedy Ann the best.

"I want braids like yours and my dolly's," Mitzi says.

"I want curly hair like yours," I tell Mitzi.

"Let's trade," Mitzi says. We giggle because we know we can't do that.

Her third favorite doll is the most dear baby doll I have ever seen. It has big blue eyes and chubby little arms and legs. She drinks from a bottle and wets her diaper. Her name is Baby.

When we get tired of playing with Mitzi's dolls, we

play "Hide-and-go-seek" and "Mother-may-I," and "Simon says." We read Mitzi's story books and play with paper dolls. We play "Snakes-and-ladders" and jacks, and color in coloring books and make shadow plays with a sheet. We swing and eat tea biscuits that are just like the ones Mama used to make. Most of all I like making up stories about Mitzi and me. I pretend we're magic and can become invisible or tiny as Tom Thumb. Elves and fairies ride away with us into the forest at midnight.

On my ninth birthday, Mitzi brings me a present in a box so big she can barely carry it.

"What is it?" I ask.

The box is wrapped in white tissue paper and has a big pink bow.

"Guess," Mitzi says.

It's not a heavy box. I hold it and rattle it and shake it. It doesn't make a sound.

"I can't guess," I tell Mitzi. What could be so light and in such a big box?

I undo the bow carefully and open the box. All I can see are big handfuls of crumpled tissue paper. I wonder if it's a joke and Mitzi's brought me an empty box.

"Keep going," Mitzi says as I take out the paper.

I take out more crumples. And then—and then—I see her. It's Baby! It's the dearest sweetest doll in the

world. I can hardly believe it. I put my hands down into the crumples and lift her up gently. She's wearing a brand new pink knitted dress with little pink booties and a pink and white bonnet. Her bottle is around her wrist with an elastic band.

"Oh!" I hold her in my arms.

"Isn't she pretty?" Mitzi says. "Mommy made the dress."

I hardly dare to ask if I can keep her.

"It's your birthday present," Mitzi says.

I want to laugh and cry at the same time. I must be the luckiest happiest girl in the whole world. I wrap a tea towel around Baby and cradle her in my arms.

"Can I really keep her?" I ask.

"Yes," Mitzi says.

11

Early one morning I waken while it's still dark outside. Obasan and Uncle are awake. Stephen is still sleeping in the bunk below me. His mouth is squished open on his pillow.

Yesterday Stephen came running home shouting that the war was over.

"We won we won we won!" he cried. He ran behind the house with both hands high in the air. His fingers were raised in the V-for-Victory sign. He pulled the flag out of the rock garden. Then up he climbed onto the shed and still higher to the roof of the house. The flag was up as high as it could go.

This morning there's no shouting. A log drops "thud" as it burns in the wood stove. The coal-oil lamp is on. Beside the lamp Baby is sitting on a tin of sardines. She's looking out the window.

Last week Mitzi and Raggedy Ann and Baby and I had a tea party in the playhouse. A chipmunk came to visit. It was the first time Baby had seen a chipmunk up close. Baby threw a temper tantrum when the chipmunk went away. She's getting quite spoiled. After she came home she wanted to sit on the table every night to watch for chipmunks. I'll have to put her to bed.

"Stephen," I whisper, leaning over the side of the bunk. I blow at his face. "Wake up," I say. Stephen keeps snoring.

"So early?" Uncle whispers as I take Baby off her sardine tin.

"You should sleep some more." Obasan is also whispering. Her long braid is dangling down her back. She hands me a piece of toast. "Sh," she says.

I give a crumb to Baby and another one for her to give to the chipmunk.

I can hear a dog barking outside. A light wind is blowing through the branches of the trees.

While I finish my toast, I notice that the other room has been changed around. The shelf is no longer against the wall and there are piles of boxes everywhere. Nomura-obasan went away months ago to stay with her daughter. But her cot is in the room again. Has she come back?

"Who's here?" I ask, standing on tiptoe.

The sleeping person is facing the other way and an arm covers the head. The arm moves then. I can see the back of the head, the straight black hair just like Daddy's.

Is it Daddy? Can it be Daddy? My hands drop with

a slap to my thighs. It is! It is!

Without turning his head, he lifts his finger and beckons.

"Good morning, my Naomi," he says.

How does he know it's me? I haven't made a sound.

He turns then, and smiles. It's my very own father. My Daddy who plays the violin and sings. My Daddy who rocks me in the rocking chair. My Daddy who munches my tea biscuits and climbs up the peach tree for the ripest peaches.

I jump over a box onto the cot and I am in his arms again—my father's arms.

His hands touch my face. I wrap my arms around his neck. The button of his pyjama top presses into my cheek. I can feel his heart's steady thump.

We are quiet as moon song. As quiet and still as resting swans. Into this quiet I fall like a lost feather returning.

We do not talk. Only Uncle says "Ah," as he swallows his tea. Obasan's butter knife makes a scrape-scrape noise on the toast. And the neighbor's dog outside barks excitedly.

Then suddenly Stephen is in the room. He stands there barefoot, rubbing the sleep from his eyes. His flutes are in his hands.

"Good morning," Daddy says.

"Dad!" Stephen howls and jumps on us. Daddy

holds us both in his arms. We rock and sway together on the little cot.

Then Stephen hands Daddy a flute and they play and play until the sky grows light.

"Whoo," Daddy says finally. "Not bad, young man."

"Not bad," Uncle adds.

Obasan turns down the coal-oil lamp. She cups her hand behind the chimney and blows out the night light. Then she gives us all pieces of toast.

12

After breakfast Daddy and Uncle talk quietly. I don't understand what they are saying, but Daddy looks sad.

"We're moving," Stephen says.

"Moving? Are we going home?"

"No."

"Why not?"

"We can't."

"Why not?"

Everyone, Stephen tells me, is going away again. But we don't know where.

I pack my Mickey Mouse and the red, white, and blue ball in a crumple of newspapers. I'm not going to pack Baby. I'll carry her myself. Stephen wraps his flutes carefully in the sleeves of his sweater.

"Are you coming with us, Daddy?" I ask. He's working more slowly and looks tired.

"No, my Button," Daddy says. "I can't. But all God's angels are going with you."

"Why can't you come with us?" I ask.

Obasan hands me a dish. She asks me to wrap it in

newspaper. I know she doesn't want me to ask questions right now. And I know Daddy has to go back to a hospital. But I want to be near him. I want him to be near us. It's not fair. I want to run away with him. But there's nowhere to go.

When it gets dark we can hear music wavering through the trees. A loudspeaker is playing ''Auld Lang Syne.'' While we're packing we hear voices and footsteps. Some people are coming towards the house.

''Good evening,'' a familiar voice calls.

Obasan opens the door and Nomura-obasan comes in bowing deeply.

''Such a busy time,'' she says. She's thinner than before and holds a cane. Behind her is an old man I know. He's wearing a suit.

''Good evening,'' he says. His voice sounds as scratchy as the screen door.

''Ah, ah,'' Nomura-obasan says when she sees Daddy. There are tears in her eyes. ''I have not seen you for such a long time.''

''Such a long time,'' the old man says. He puts his shaking hand on Daddy's shoulder.

Obasan pushes the boxes aside as the minister comes in. He takes a long black gown and a shorter white gown out of his black bag.

''Let us pray,'' he says as people kneel on the floor. ''This is the last supper.''

The old man tries to kneel but he can't. He leans on his stick.

My eyes are supposed to be closed. But I'm peeking at everyone's feet. The minister's boots rock back and forth as he prays. His words sound like the rustling leaves in the fall when the wind blows them about.

The old man's false teeth make a clacking sound. His voice wheezes as he stumbles to keep up to the others. Nomura-obasan can't keep up either. She's shaking so much that Obasan has to hold her.

When all the prayers are finished, everyone sings a goodbye song.

> *"Till we meet*
> *Till we meet*
> *God be with us*
> *Till we meet again."*

Daddy's eyes are closed. He's trying to sing too but sometimes he stops. The old man's singing is out of tune with the others.

"Once more," the minister says. "Let us sing again."

The voices fill the tiny room. I feel I will always hear them singing—the dear old man and Nomura-obasan and the rest of us.

When the song ends for the third time, Obasan

holds the old man's bony hands.

"Let us meet again some day," she says.

Nomura-obasan takes a handkerchief from her sleeve. She holds it over her trembling face. The minister puts his hand on her back.

"We will trust in God," he says.

"There is a time for crying," the old man says in his wavery voice. "Someday the time for laughing will come."

"Yes, that must be so," the minister says. "We will meet again."

He puts a hand on Stephen's head. "Be a great musician like your father," he says. Next he turns to me. "Be sturdy." He bows to everyone. Then he is gone, trotting rapidly down the path to the next waiting group.

13

Daddy goes back to the hospital the next day. After a few more days the time comes for us to leave.

"We have to say goodbye to Mitzi now," I tell Baby.

Obasan and I take a cake with us. When Obasan visits anyone, she always takes a present.

"Goodbye, Baby," Mitzi says, kissing her.

"Goodbye, Raggedy Ann," I say to Mitzi's doll.

We say goodbye to all the dolls and the bunnies, Patsy and Gruff. There are new little baby bunnies and we say goodbye to each one. We say goodbye to the swing and the playhouse. We start giggling as we say goodbye to the tea set and the table and the doll dresses and the coloring books and the chipmunk who isn't even there.

"God bless you all," Mitzi's mother says as we leave.

At the train station there are boxes and luggage and hundreds and hundreds of people. It's like the day when we first came to Slocan. The black noisy train clangs its bells and hisses back and forth.

Some of the children we used to meet at the bath and at school are here. Above the noisy crowd, the

scratchy loudspeaker plays "Auld Lang Syne." Kenji's older brother is in front of Stephen with a black bag over his shoulder. I can't see Kenji.

Hoo-oot! goes the train. One by one we move along. We're like a giant caterpillar. Uncle is behind me and lifts me up onto the steps.

Inside, it's just like three years ago except that Uncle is with us. We sit in two seats facing each other. Kenji's brother is ahead and there are some others I recognize. The minister's on this train too.

People outside are waving and waving. Some are crying. It's so sad to have to say goodbye. I'm remembering the time at the boat in Vancouver when Mama went away. I wonder where she is now. And where, I wonder, is my old doll.

The train shudders and starts to move. I press my face against the window. Stephen and Uncle stand up to wave as we pull away from the station. Then almost right away we're into the thick trees and can no longer see anyone.

It's goodbye to the mountains, the lake, to Rough Lock Bill, the school, the bathhouse. "Goodbye everything. Goodbye everyone," I whisper to the train window.

We enter a high tunnel as we race along. Clackity clack, clackity clack, clackity clack. "So long, Slocan."

14

Finally we come to our new home. What a dusty, lonely place. The air here is angry and hits out suddenly like a wild man. It blows dust and dirt into your eyes and your hair. You turn around and turn around and squeeze your eyes shut. But you can't escape. The flat brown earth stretches on and on till it meets the sky. Dried bunches of scratchy weeds tumble along the fields and roads. They get stuck on the miles and miles of barbed-wire fences. No trees can stand this awful place.

Our hut is even smaller than the one in Slocan. There's just one room. Out of one window we can see the huge farm machines. They look like skeletons of dinosaurs. From the other window we can see the straight road with the ditch beside it.

Obasan puts rags and newspapers around the bottom of the door and the windows. She's trying to keep out the dust and the flies. But they keep coming in anyway. In summer the windows are covered with them. The horrible flies walk on your arms with their sticky hairy feet. They stick in your hair and land in your food. Ugh! Why don't they just go away?

We don't have a bathhouse here. Our bath is a round tub. Getting water is such hard work, especially in winter. We put on our boots and coats, and out we go with our buckets. The hole always gets frozen over and Uncle has to chop it open with a long-handled axe. I can hardly lift the heavy pails. The water sometimes spills down my boots and my feet get itchy and bumpy and red.

After we all take our baths, Obasan washes the clothes in the same water. They hang outside in the icy wind, stiff as cardboard. It's so cold your face stings and your eyelids freeze.

I hate it here. I hate it so much that I want to run away. So does Stephen.

"Why can't we go away?" I ask Uncle. "Even if we can't go back to our first house, can't we go back to Slocan?"

"Someday. Maybe someday," Uncle says. But "someday" never comes.

In the spring we have to work, work, work. The field stretches on forever and is full of rows of plants. All day long we hoe the weeds. It gets so hot it feels like an oven. We're gingerbread cookies baking to bits.

Sometimes I get sick from the heat and lie down in the dirt. Then Uncle comes running across the field. He carries me to the root cellar or to the ditch water. The root cellar is cool, but the rotten potatoes smell

horrible. I'd rather sit under the bridge in the muddy water.

"Careful," Uncle says as he helps me into the brown water. The thistles growing on the ditch bank sting your feet.

The school in Granton is different from the one in Slocan. Most of the children don't have black hair like Stephen and me. And they don't have to stay home to work like us either. Only the children like Stephen and me have to work. The teachers send us our homework to do at night.

In harvest time, Obasan wraps rags around all our wrists. She says it helps to lift the heavy beets. But I don't like the rags. I don't like us looking dirty and ragged and ugly in the dusty field. I have to wear Stephen's old clothes. None of my dresses and skirts fit anymore and I don't have pretty clothes. For school, Obasan fixes her old dresses to fit me. But they don't fit. She says they are beautiful silk. But I hate them. I want store-bought dresses like the other girls.

I write long letters to Daddy in the hospital. He can't come to be with us, Stephen says, until the doctor says he can work. Daddy always sends music to Stephen.

Stephen saves Daddy's sheets of music and ties them all together with shoelaces. One evening he's finally finished helping Uncle with the irrigating job.

He takes out all Daddy's music and kneels to get the flute from under the bunk bed.

"Uncle!" Stephen cries out. He holds the flute up for Uncle to see. There is a long crack all the way down the side.

"Ah, the air is too dry," Uncle says sadly.

Stephen and Uncle try tying it together and taping it. But when he plays it, it just sounds windy. He tries and tries until Uncle says finally, "We can't fix it."

Stephen takes Daddy's music and a flashlight and runs out of the house. We can hear him pushing the bike away from the house.

"Where are you going?" Obasan calls from the door. But Stephen doesn't answer. From the window we can see the light from his flashlight bouncing up and down as he goes down the road. It's the first time Stephen has gone to town on his bike at night.

Bedtime passes and still Stephen doesn't come home. He doesn't come home all night long. Early the next morning, as Uncle is getting ready to go to look for him, we can see Stephen riding his bike. He looks like a dot on the road.

"Where were you?" Obasan asks. She's been sitting up all night. She left the coal-oil lamp on in the window.

Stephen just shrugs his shoulders. He doesn't want to talk.

Uncle doesn't say anything.

The next day, when we're hoeing, Stephen tells me he went to the United Church in Granton. A window was open. He climbed in and felt around till he found the piano. He also found a blanket and covered the piano so it would be quieter. He was afraid he might get caught. Then he played Daddy's music until the flashlight batteries died.

"Want to hear Daddy's songs?" he asks. The tunes he whistles are so happy they make me want to dance.

Stephen says when the flashlight went dead, he took the blanket off the piano and fell asleep.

"When I grow up," Stephen says, "I'm going to have a piano and a violin and another flute. I really am, Naomi. And a trumpet too."

"I believe you, Stephen," I say as I whack out the weeds.

15

Across the field from our hut is a place where the cows are. I bring my baby doll and my Mickey Mouse to play here. There's a dead tree and a swampy slough of water with bulrushes and bushes around it. And what a lot of creatures there are! The longer I watch, the more I can see. There are wriggly tadpoles, dragonflies, water spiders, mosquito babies, jelly eggs, frogs, toads, busy ants. If I make even a little noise, all the swamp sounds stop.

I pretend we're back in the mountains. But it isn't like the mountains at all.

"Wouldn't it be nice if Mitzi was here?" I whisper. "And Raggedy Ann too?" I remember I used to pretend that my Japanese doll could talk. But I don't play like that anymore.

One Saturday, after supper, I'm sitting by myself at the swamp. The frogs and toads are croaking and breeping us usual. Once in a while a meadowlark sings. All along the edge of the sky, the clouds are changing colors. It's quiet and still, yet the world looks like it's on fire. I remember Mama used to say that a match was safe if you could blow it out. But what if the whole world was on fire? How could you blow that out?

I'm staring right into the fierce red and purple sunset. Gradually I notice that three people are coming down the road. They seem to be walking in the middle of a huge fire.

In the Bible there's a story of an angel. The angel kept three men safe when they walked in a fiery furnace. Who are these three, I wonder. And where is their angel? I ought to be safe, I guess. Daddy said that all God's angels were going to be with me.

As they get closer I recognize Stephen first. He's pushing his bike and whistling. He's trying to copy the meadowlark. Stephen must have an angel that flies through the air gathering music. Uncle is walking on the other side, carrying a bucket, and the minister is in the middle. He's pushing his bike too.

I stand up and wave to them. Uncle and the minister wave back.

"Hello, Naomi," the minister calls.

All the frogs and toads suddenly stop their noise. Plip ploop, they go, as they dive for safety.

"Coming with us?" Stephen asks.

The minister says that he saw some beautiful mushrooms beside the road. Uncle and the minister know which mushrooms are safe to eat.

"It wasn't far from here," the minister says as we

walk along. He slows down as we come to a sandy stretch. There are some wild rose bushes growing on the other side of the ditch.

"Ah," the minister says, putting his bike down. "See how generous the earth is to us." And there, along the whole stretch of sand past the roses, is an enormous crop of mushrooms.

"Such treasures," the minister says as he helps us fill Uncle's pail. "These are such treasures." When the pail is full Stephen pulls off a rose and puts it on top of the mushrooms. There are still lots of mushrooms left to pick.

After a while the minister rides off down the road in the graying light. He's taking a bagful of mushrooms to another family five miles away. The minister always brings news or gifts. Tonight, Stephen says, he brought a letter for me.

"For me? Who from?" I ask.

"Someone in Slocan," Stephen says.

I wonder who it can be.

By the time we get back to the hut, the stars are bright in the sky.

I don't recognize the handwriting on the envelope at all.

There's a white piece of cardboard wrapped inside the letter. I read the letter first. It isn't very long. It goes like this:

Dear Naomi,
I am fine. How are you? It is raining today.
My mother is going to send my letter to your
minister. If you get this, please write to me
right away.

Goodbye,
Your friend,
Mitzi

P.S. Send me a card like mine and we'll be
you know what.

Super P.S. My mother says to say hello to
everyone.

Extra Super P.S. How is Baby? Smudge the
chipmunk says to say hello.

Extra extra etc. etc. Patsy and Gruff have
more babies. I wish you could have one.

That's the end of the letter. I can guess what's on the card. I'll bet anything I'm right. But I'm not going to look yet. I feel like shouting but I giggle instead. It makes me giggle to think of Smudge saying hello. It makes me giggle just thinking of how much Mitzi and I giggled and giggled.

"Who wrote?" Stephen asks.

I show him the letter. But I won't show anyone the

card. I hide under the blanket with a flashlight.

I was right! There's a small blotchy mark on the card beside her name. "THIS IS BLOOD" it says under the mark. Above the mark is a verse.

WE TWO ARE A SISTERHOOD
WE SEAL THIS SECRET
WITH OUR BLOOD

We're blood sisters! I feel like jumping out of the covers. But I mustn't let anyone know. It's a secret. A blood sister is forever. If we tell anyone, the magic will be broken and our secret codes will be destroyed. If one of us gets caught in a war, we have the power to rescue the other.

We talked about it in our hide-out in the woods. But we didn't have a needle. I was glad we didn't. Poking your finger hurts. We printed our promise on a white piece of birch bark. I wonder if it's still there under the stone where Mitzi and I once buried a dead bird.

I hide Mitzi's card under my pillow. Stephen is sitting at the table biting his pencil. He's doing his homework. Obasan is washing the mushrooms.

"So big," she says. "Such fat mushrooms."

Uncle has put the pink rose in a glass vase. It's on the table in front of Stephen.

No one is watching as I hunt for a needle in Obasan's sewing box. There's one with a long black thread on it.

I squinch my face up and jab my finger. Oo! A tiny dot of blood comes out when I squeeze hard.

I take Mitzi's card from under the pillow. Bap! There it is now. I've made a blood blotch too. I take out my pencil and sign my name beside my blood. It's done! Now, at this moment, our sisterhood has begun. Tomorrow I'm going to the swamp to write Mitzi a letter.

16

Where am I? What is this place?

Daddy? Daddy, don't go yet.

Wait. This can't be right. "Daddy!" I'm whispering out loud. I really am awake now. But wasn't I awake just a minute ago?

There's a light glow in the room. I don't know where it's coming from. It isn't grayish white like the light from the moon. It's more yellowy. My head feels funny, as if there's a wind rushing around in it.

Was I just having a dream? It didn't seem like a dream. First, Mama was here in this very room. So was Daddy. I can still feel them here. Mama was inside the wild rose in the vase and Daddy came out of the mushrooms. They were singing the daffodil song.

> "How did you, Miss Daffodilly,
> Get your pretty dress?
> Is it made of gold and sunshine?
> Yes, child. Yes."

I was so glad to see them that I sat up in bed and put out my arms. I was little again.

"Don't go away, Mama," I cried. "Daddy, don't go away."

But as I reached out to them they began to fade and disappear.

"A match is safe if you can blow it out," Mama said.

And suddenly the rose was burning. I couldn't get out of bed. I couldn't cry out. Somehow, then, I knew I was dreaming. I tried and tried to wake up, but I couldn't. I was falling and falling and I couldn't see and I was dizzy. It was almost like the time in the lake in Slocan with Rough Lock Bill.

Before I could land I broke through. I was awake in my bed. My eyes were still closed. I knew I was really awake, but I could see the room with my eyes closed. Daddy was right here. He sat beside me on the bed. I thought, "Isn't this funny. I can see Daddy with my eyes closed."

"God's angels are always with you," he said. Then he hugged me so tight I can still feel it. I really did know he was here.

I said, "Daddy." And then my eyes opened like they are now. I know I know I know he's here. So is Mama. Even if I can't see them. Everything I see is a kind of fuzzy golden dream.

The mushrooms are washed and sitting in a bowl. And the wild rose is on the table. I tiptoe over to smell the rose. It smells like Mama's perfume.

Uncle turns his head as I climb off the bench. "So early," he whispers.

"I had a dream," I whisper back. I know that's the way I have to tell him about it.

Uncle sits up listening carefully and nodding his head. "She was singing in the rose?" he asks. "In the rose? Right now? And fire?"

I nod.

Uncle lowers his eyes. "Are they still here, Naomi?" he asks softly.

Now that I'm talking to Uncle, I'm not sure anymore. I look at the rose and it just looks like a rose.

"I don't know, Uncle."

Uncle stares and stares at the rose. Finally, he whispers, "Pack some food, Naomi. We have somewhere to go."

Obasan opens her eyes. "Why are you awake?" she asks.

Uncle waves his hand. "Sh. Sh," he says. "It's too early to be up."

I don't know where Uncle and I are going. I put some peanut butter sandwiches and two apples in a bag as quietly as I can. Uncle fills a bottle with grape juice from a can.

The sky is gradually getting lighter as we start walking across the field. We pass the swamp and keep on going. When we get to the mushrooms, Uncle stops.

"Here," Uncle says. "This is where we'll sit."

The ditch is almost dry except for a small trickle of water at the bottom. Uncle squats on the slope of the ditch beside the mushrooms. The long shadow of the rose bushes reaches across the road. I squat beside Uncle in the speckled morning shadow.

Close by, a meadowlark is singing. And from far across the fields, at another farm, a rooster is crowing. We sit and sit as the earth grows gradually warmer and the shadows shrink.

I grow tired just sitting here in the sand.

"I'm hungry, Uncle," I say at last.

He doesn't move. He's staring straight ahead. He's looking right through the roses. I wonder if he can hear Mama singing there. Or maybe he can see Daddy in the mushroom patch. I can't hear them or see them at all now.

"Do you want a sandwich, Uncle?" I ask.

He sighs a deep sigh and looks at the sky. "No one knows the ways of the universe," he says quietly. "No one knows. No one knows."

We sit eating our sandwiches and drinking from the bottle of juice. Ants come to take away the crumbs.

When we finish eating, Uncle gathers more mushrooms. The paper bag is almost full when we see the minister. He's riding on his bike towards us.

"Hi," I call, waving.

"Good morning," he says as he comes closer. "You're up early."

"Ah," Uncle says, straightening his back. "We came to smell the roses."

"Good, good," the minister says, nodding.

As we walk back down the road, Uncle and the minister talk about mushrooms and roses.

"Mm," the minister says thoughtfully, when Uncle tells him about my dream. "And the rose was on fire?" He shakes his head. "There are ways to walk safely when the world is in flame. There are ways to blow the fire out."

I wonder what the minister means.

"There is a straight road to walk," he says. He seems to be talking in riddles.

"What's the road's name?" I ask.

"Everyone has some treasures and our road goes to where our treasures are," the minister says. "The names of our roads are the names of our treasures."

Yesterday the minister said the mushrooms were treasures. This must be Mushroom Road we're on. But really, this road doesn't have a name at all. The one beside our hut is called Number Seven Road. What a boring name.

Uncle smiles. "The best treasures are friends," he says.

The minister nods. "Yes. A treasure that we keep here." He taps his chest. "Friendship Road is an important highway."

I'm thinking that Mitzi is a good name for a road. But this road we're on, I'll call Mushroom Patch Road. And the swamp will be Mitzi Meadows. I'll make a sign and put it on the tree where I sit.

I start skipping down Mushroom Patch Road towards Mitzi Meadows. Uncle and the minister are still talking in riddles behind me. But I feel as if Mitzi and Mama and Daddy were walking along with us in the golden sunshine.

When we get near the hut, we can hear Obasan singing the daffodil song. I've never heard her sing it before. I didn't know she knew it.

"That's what Mama and Daddy were singing this morning," I tell Uncle and the minister.

"Yes," the minister says, smiling and nodding. "The world is full of signs. We have to know how to read them."

They're going to keep talking in those riddles that adults like. But I'm going to get Baby a pencil and paper to take to Mitzi Meadows. Mitzi and I have a lot of codes to work out. It's going to take me all day.